TECHNOLOGY IN EDUCATION AND BUSINESS

TECHNOLOGY IN EDUCATION AND BUSINESS

Myths, Issues, Ethics, and Money

Edited by

JJ Asongu
Adjunct Professor at University of Phoenix &
Chattahoochee Technical College, Marietta, GA.

&

Dana Lundell
Assistant Professor, College of Education
Argosy University-Twin Cities

Greenview Publishing Company, Lawrenceville, GA 30045
In collaboration with Pinnacle Seminar 2007, Argosy University

Library of Congress Control Number: 2007940193
ISBN: 978-0-9797976-2-0

Greenview Publishing Company
455 Grayson Highway
Suite 111-118
Lawrenceville, GA 30045
http://www.greenviewpress.com
Email: info@greenviewpress.com
Tel: 770-780-2739
Fax: 770-513-3786

Printed in the United States of America

Table of Contents

INTRODUCTION

1

Pinnacle Seminar at Argosy University-Twin Cities

Dana Lundell[1]

This mini-publication is a sample of the collective efforts of a group of 8 students participating in Argosy University's Pinnacle Seminar (PS2007) in the first fall session of 2007. The course topic was "Technology and Ethics for Education and Business Leaders." Students were required to take an online course with a 3-day residency component where they met together with their national peers who were also taking sections of the class.

The seminar's intention is to allow students to explore a theme deeply and apply it to their own workplace setting, such as a school or business. A core component of the class was a project called "Field Experience," which emphasized a

---

[1] Dana Lundell, Ph.D., is an Assistant Professor in the College of Education at Argosy University-Twin Cities. Formerly she was Director of the Center for Research on Developmental Education and Urban Literacy at the University of Minnesota-Twin Cities. She has also edited and authored publications in the fields of postsecondary developmental education, multicultural education, and graduate student development.

topic area, such as issues, myths, ethics, and funding related to technology. For many students, the topic was brand new, and the writing requirements were intensive in terms of promoting long-term exploration of a topic with the opportunity for feedback and multiple revisions.

I was the faculty member assigned to this group of students. They were all from the Atlanta campus, and I was a new faculty member in the Education department at the Twin Cities campus. We met online, and our relationship blossomed from there when we finally met in person at the seminar's residency in Minnesota in October 2007. I worked diligently with these writers to help them accomplish the goal of preparing a short research-based paper on this theme in APA style.

They went through several drafts of review and feedback; at the seminar's conclusion, this mini publication was the result. I am proud of this group of authors who learned to open themselves up to sometimes difficult writing feedback from their professor. I learned that this group of students was ultimately very diligent and sought improvement and intellectual growth in every arena of their graduate student lives.

I thank these students--Januarius ("JJ"), Kimberly B., Beverly, Sharnelius, Tamika, Mary "Frances," Precious, and Kimberly S.--for their efforts and willingness to work hard in this class. My entry into the world of online teaching was interesting and enjoyable as a result. Thanks also to JJ for his interest in making this publication into something more useful to others, and we are all grateful for his efforts.

Through working with these graduate students, I learned and recommend that faculty members can and should offer a great deal of writing feedback to students when possible. It takes time, but it will pay off in their lives as writers in their future professions, whether it is for their dissertations or for their work as educators and business professionals. This mini-publication was only a sampling of their work as a whole, but it allowed us to experience the process of deeper writing and revision in a short piece of academic writing.

The future is certainly brighter with this group of students out there, all working toward creating more innovative and ethical school systems and business settings through the implementation of new technologies. Thanks, everyone, for your great work!

ISSUES

2

Benefits of Technology in the School Systems

Beverly Barnes[2]

Schools systems must implement a plan for integrating technology into the curriculum in order to realize its benefits. Not only should technology be implemented for curricular purposes, but also for safety and security, financial benefits, and for empowering youth to cross the digital divide and become involved with global issues as future leaders. According to North Central Regional Educational Laboratory (1995) an effective technology plan is based on the shared vision of educators, parents, community members, and business leaders who have technological expertise. It ensures that technology strengthens existing curricula and supports meaningful, engaged learning for all students.

Given that technology continues to play an important role in modern industrial society, integrating technology into

---

2 Beverly Barnes is looking forward to earning a Doctorate in Instructional Leadership at Argosy University, Atlanta. Currently employed by the DeKalb County School System, she teaches Social Studies. Her ambition is to become an expert in designing and organizing educational institutions that can move progressively forward through the 21st century.

the schools will help prepare students to succeed in a rapidly changing world. It is the responsibility of schools to prepare students with the new technology that is being introduced in our growing industrialized society. School systems must continue to become knowledgeable in advanced technology, thus integrating it into the schools' curricula.

According to the New York State Education Department (2007) technology is human innovation in action. It involves the generation of knowledge and processes to develop systems that solve problems and extend human capabilities. Educational technology is the execution of multimedia or audio-visual aids as a tool to improve the teaching and learning process. A school system is an organization that offers education for children pre-kindergarten through twelfth grade.

Technology Integration for All Stakeholders

Educators, parents, and community members are more inclined to become supportive of technology if they are able to see evidence of its importance and effectiveness in helping students learn. The plan for integrating technology into the school is based on the school or district's educational vision and is part of an overall school-improvement plan. This technology plan aims to improve student learning, to help students perform dependable jobs, and to help students learn skills that will prepare them for future careers. Professional development and support for staff is provided on an ongoing basis. Assessment plans should be in place to ensure the technology plan produces the desired outcomes.

A Resource for the Community

Technology is beneficial as a resource for the community. According to Family Technology Resource Center (Bouie, 2007) one initiative is a simple, yet highly effective program to address the difficulties those without technology skills face in a booming digital economy. FTRC features community-supported learning environments for all ages, educational levels, and economic status.

Family Technology Resource Center creates multi-generational community-learning centers. Students stay after school to use courseware and software designed to improve academic achievement, communicate with global classmates, participate in activities via the Internet, and develop their own personal skills. Parents and other adult community members have access to educational technology and participate in instructional activities together, including basic literacy, GED completion, English as a second language, and family school-community activities.

Distance Learning

School systems also offer convenient and intensive learning through technology. Online Academy, a distance learning program for high school students, has become a popular way of preparing students for the education delivery model of the future, and it ensures that students have the technology competency skills needed to compete in a global society. This distance learning program is a practical option for all students for enrichment and enhancement.

Administrators, Faculty, Parents, and Uses of Data

Over the last two decades, programs used for storing and retrieving data, have improved drastically for educational leaders and teachers, such as Testatrix and Smart Web. According to DeKalb County School System (2007) TestTrax is an Internet Web-based methodology that allows tracking, analysis, and reporting of standardized student test scores. It crosses multiple years, allowing the establishment of growth and performance trends for a single student, as well as for classroom, school, and district groupings of students. Along with TestTrax is Smart Web, a program that diminishes teacher work load, eliminates floppy disks, and also helps students keep up with their grade point average. This program also gives parents access to grades, attendance, and homework assignments.

Conclusion

Integrating innovative and new technology in the school systems will ultimately pay great dividends in the end. The whole community, at large, should have a vested interest in the implementation of this fast growing resource. School systems can no longer afford to stand idle while the world is drastically evolving into such a modern industrial society. Education is the key to preparing our youth for success in this evolving world.

References

Bouie, E. (1996). *Family technology resource center.* Retrieved September 22, 2007, from http://www.ftrc.org/mainframe.asp

Decal County School System. (2002). *In the media.* Retrieved September 22, 2007, from http://www.smartwebtk.com/inthemedia.htm

DeKalb County School System. (2007). *Our commitment to you.* Retrieved September 22, 2007, from http://fc.dekalb.k12.ga.us/~ctdola/DOLA/Mission.htm

New York State Education Department. (2007). *Definition of technology terms.* September 14, 2007, from http://www.emsc.nysed.gov/technology/nclb/definition.htm

North Center Regional Education Laboratory. (1995). *Developing a school or district technology plan.* Retrieved September 22, 2007, from http://www.ncrel.org/sdrs/areas/issues/methods/technlgy/te300/htm

System, D. C. (2002, March 4). *In the media.* Retrieved September 22, 2007, from http://www.smartwebtk.com/inthemedia.htm

TestTrax. (2007). *TestTrax: Standardized test analysis.* Retrieved September 22, 2007, from https://testtrax.dekalb.k12.ga.us/tt-login.asp

3

Issues Relating to Ethics and Technology: Cellular Phone Usage of Student for Cheating

Sharnelius Colbert[3]

Major issues of concern in school systems around the United States include lack of student motivation, low graduation rates, illegal weapons on school campus, and the use of cellular phones at school. The issue of cellular phone usage has recently gained heightened attention.

Most educators believe cellular phone usage should not be permitted during the school day and support the confiscation of cellular phones if they are visible at school. Meanwhile, some parents feel that students should be allowed to carry cellular phones for the purpose of communication in the event of an emergency.

---

[3] Sharnelius Colbert is a doctoral student in Educational Leadership (K-12) at Argosy University, Atlanta. A native of Florida, she now resides in the Atlanta, GA, area, where she currently works as a high school guidance counselor in Clayton County. She has been employed in the counseling profession for six years and enjoys working with students. Her goal is to become an administrator at the secondary level.

The debate as to whether cellular phones usage should or should not be allowed in schools continues. Educators and parents both have their reasoning for their positions on this issue.

Research shows that the history of electronic device usage in schools can be traced back to the 1980s (St. Gerard, 2006). During this time many schools did not allow students to carry electronic devices such as pagers or beepers. These devices were banned because of their association with the drug trade (St. Gerard, 2006). St. Gerard (2006) asserts that because of tragedies such as the shootings at Columbine High School and the terrorist attack on the New York Trade Center and Washington, DC, many schools lifted their ban and allowed electronic devices into schools after seeing how beneficial they are as a means of communication. In Cleveland, Ohio, David Root, principal of Rocky River Middle School decided that after such tragedies students at his school could bring cell phones to school and could use them before and after school. Students at Lewistown Junior High School are also allowed to bring cell phones to school (St. Gerard, 2006). These two schools are two of many schools around the United States that have changed their stance on students' use of electronic communication devices in schools.

With the advent of students being allowed to carry electronic devices to schools educators are finding themselves having to deal with matters such as students cheating on examinations, hearing irritating ringbones, text messaging during class, and pictures being taken to post on websites such as MySpace (Brinkley, 2006; St. Gerard, 2006; Yaqoob, 2006). A report released by the Qualifications and Curriculum Authority (QCA) reported that more than 1,000 students were caught with cellular phones that were used for cheating. A report conducted by QCA 1995 revealed that one in four cases of 4,500 cases of students penalized for malpractice during examinations was for smuggling cellular phones in the exam room for the purpose of cheating ("Education Matters," 2006).

The use of cellular phones in schools has become such a problem that school districts are implementing school policies to regulate cellular phone usage. In Baker County personal telephones may not be used by students until before and after school. If telephones are used they are confiscated and returned only to a parent or guardian. In Duval County if cellular phones are used during the school day, a parent is contacted and the student receives one to three days' suspension for the first offense. In other counties in Jacksonville, Florida, students may receive detentions, or in-school or out of school suspension (Tseng, 2005). In Duval County 237 students were suspended for violating cellular phone policies (Tseng, 2005).

Technology continues to change with the invention of new equipment. The cellular phone is one piece of technology that many parents are finding to be a necessity for their students as a form of communication. The purpose of cellular phones originally was to communicate. This purpose is continuously evolving. Currently the cellular phone is being used for many recreational activities. Students are finding cellular phone usage to be a social avenue, and more and more students are receiving cellular phones.

The use of the cellular phones has been taken to another level. Students are taking pictures in class and sending the pictures to other students and even posting the pictures on the Internet. Some of these pictures are being taken without the person's permission. Fights, classroom disruptions, and many other school activities are being recorded on cellular phones. The use of cellular phones to cheat during exams is another cause for concern as such actions are generally considered unethical at the school level and by parents. This concern is on the rise as it is being used at elementary schools, middle schools, high schools, and at the college level. The debate continues as to whether cellular phone usage is appropriate during school.

References

Blake, A. & Pascoe, C. (2006). Mobile phone cheats on increase in school exams. *Western Mail.*

Blinkley, M. (2006). Mobile phones. *Western Mail.*

Education matters: The cheating exam teenagers who 'phone a friend.' (2006, September 25). *The Birminghman Post & Mail*, p. 5.

St. Gerard. (2006). Updating policy on latest risks for students with cell phones in the school. *Education Digest*, 72 (4).

Tseng, N.-H. (2005). *Student cell phone use taboo in area schools.* Retrieved September 8, from http://cgi.jacksonville.com/cgi-bin/printit.cgi?story

Yaqoob, T. (2006). Mobiles' should be blocked to stop the exam cheating. *The DailyMail.*

4

A Single Gender Academy's Response to Technology and the Related Ethical Issues

Tamika Hibbert[4]

On March 3, 2004, Secretary Spellings announced more choices in single-sex education which provided local school districts with the flexibility to create single-gender classrooms and schools (U.S. Department of Education, 2004). The Atlanta Public School Superintendent, Dr. Beverley Hall, made the decision to pilot two single-gender academies at Benjamin S. Carson Honors Preparatory School. According to Hall (2006), the district

---

[4] Tamika Hibbert is currently the School Counselor for an all-girl Single Gender Academy in the Atlanta Public Schools System. In addition she an Adjunct Professor at a community college, where she teaches courses in Education, Counseling, and Psychology. She holds a Bachelors in Psychology, a Masters in Education in both School Counseling and School Administration, and is currently a doctoral student at the College of Education, Argosy University, Atlanta. She relocated to Georgia from New York, where she had been a school vice principal.

selected Carson based on the challenging demographics of the community and the need to improve student performance. As a result, the Coretta Scott King Young Women's Leadership Academy opened on August 13, 2007. One hundred and eighty six female students currently attend the academy, and ethical issues continuously arise when technology is incorporated into the instructional activities.

Definition of Terms

According to Nelson, Palonsky, and McCarthy (2007) the term "issues" in the context of education reflects basic human disagreements as it relates to ideological differences in politics, economics, and the social values that underline the battles over schools. Some of the major issues that evolve in education include the debates about how education should be controlled, financed, organized, conducted, and evaluated. These issues influence the policies and procedures that are carried out in schools all over the nation.

Wikipedia (2007) defines "ethics" as a general term for what is often described as the science of morality. Kennedy and Krumine (2001) identify ethics as the study of fundamental principles that defines values and determines moral duty and obligation. Beech and Chadwick (2004) describe ethics as the principles or assumptions determining the way individuals or organizations ought to conduct themselves.

A single gender is a population that is comprised of one sexual category. In a single-gender school, the entire student population across all grade levels encompasses one gender (Hall, 2006). The teachers deliver instruction that is based on the gender learning differences identified by the Guarian Institute.

Scenario of a Core Issue

Every classroom at the Coretta Scott King Young Women's Leadership Academy has a minimum of six desktop computers. All of the computers are connected to the Internet, which gives the students the opportunity to conduct assignments and research topics throughout the class lessons. Although students are given permission to use the computers, some students use the technological services inappropriately. Students are using the Internet to log on to websites such as MySpace.com or YouTube.com. However, login on to such websites is a violation of the school district's general regulations for Internet access.

Staff Member's Role

According to the Atlanta Public Schools Acceptable Use Policy (1999), the district cannot guarantee the accuracy of information or the appropriateness of any material that a student may encounter. Therefore, certified staff shall supervise students while using online services at the school site and may ask instructional assistants and students to assist in this supervision. The principal or designee shall oversee the maintenance of each school's technological resources and may establish guidelines and limits on the student's use (Atlanta Public Schools, 1999). These practices are expected to increase the likelihood of students using technology in an ethical manner.

Student Behaviors

Students and their parents or guardians are given the Atlanta Public School Acceptable Use Agreement Form, which specifies the user obligations and responsibilities when using Internet services. Students are supposed to keep their account information private and only log into the system under their own account number. The students must solely use the system for educationally related purposes. The Atlanta Public Schools Acceptable Use Policy (1999) also indicates that students are prohibited from accessing, posting, submitting, publishing, or displaying harmful subjects or material that is threatening obscene, disruptive, sexually explicit, or anything that could be construed as harassment. Inappropriate use may result in a cancellation of the student's access to the system privileges. In addition, students who fail to abide by the district's rules shall be subjected to disciplinary action, revocation of the user account, and the legal action as appropriate. Unfortunately at the Coretta Scott King Young Women's Leadership Academy some students' access to the computers have been suspended.

Conclusion

It is very important that, as globalization increases, students are provided with opportunities to incorporate technology into their daily practices at school (Stewart, 2007). However, it is important that schools are proactive versus reactive when addressing the ethical issues that arises regarding students' usage of technology. One strategy that staff members at the Coretta Scott King Young Women's Leadership Academy can implement to prevent any dilemmas from occurring is ensuring that all students using the technology systems receive training surrounding proper usage. Counselors and advisors could also conduct classroom guidance sessions that address true scenarios about the consequences that are associated with improper technology practices. This may increase the goal of incorporating technology into the schools in a positive manner.

References

Atlanta Public Schools. (1999). *Acceptable use policy.* Retrieved September 21, 2007, from http://www.morrisbrandon.com/forms/acceptableusepolicy.pdf?

Beech, J., & Chadwick, S. (2004). *The general sports business webpage.* Retrieved September 23, 2007, from http://www.booksites.net/download/chadwickbeech/Glossary.htm

Hall, B. L. (2006). *Concept for the establishment of two single-gender secondary schools.* Retrieved April 17, 2007, from http://www.atlanta.k12.ga.us/content/schools/SGpresentationweb0607.ppt

Kennedy, B. K., & Krumine, L. A. (2001, January 31). *Penn state lectures on the frontiers of science.* Retrieved September 23, 2007, from http://www.science.psu.edu/alert/frontiers/Glossary1-2001.htm

Nelson, J. L., Palonsky, S. B., & McCarthy, M. R. (2007). *Critical issues in education.* New York City: McGraw-Hill.

Stewart, V. (2007). Becoming citizens of the world. *Educational Leadership.* 64(7), 8-14.

U.S. Department of Education. (2004). *Nondiscrimination on the basis of sex in education programs or activities receiving federal financial assistance.* Retrieved March 20, 2007, from http://www.ed.gov/print/policy/rights/reg/ocr/t9-noi-ss.html

Wikipedia. (2007). *Ethics.* Retrieved September 21, 2007, from http://en.wikipedia.org/w/index.php?title=Ethics&oldid=159375944

MYTHS

5

The Underlying Myths of Today's Technology-Driven World

Kimberly Badger[5]

Myths and misconceptions about technology plague the worlds of business and education. Being empowered by technology has divided the beliefs regarding productivity, communication, and their possible enhancements or obstacles. With the infatuation of the latest gadgets and program software, it is not a surprise that people have considered and are now conducting school and work from home.

Those who are technology savvy advocate the employment of technology in these ways, while traditionalists dissuade these actions for reasons of their

[5] Kimberly Badger is from Barnwell, SC, and she is pursuing a doctorate in Instructional Leadership (Higher Education) at Argosy University, Atlanta. Although her goal was to be a college Mathematics professor, she is currently excited with teaching Algebra in Aiken County, SC. She is interested in educational research and still intends to be a university professor.

own. In order to address this controversy, myths about technology as they relate to school and work must be identified and given the proper response, which is the truth about the side effects of technology.

The Education World: Myths about the Digital Classroom

First, the option of the digital classroom is increasing around the world, and so is its role as the topic of debate. People tend to disagree about whether its efficiency overpowers its impersonality or vice versa. One myth concerning digital classrooms is that using the technology and Internet increases productivity and saves time and money. One investigating professor of a qualitative study had a different point of view, deeming the digital classroom as more time-consuming with the obligation to post feedback in a timely manner. As stated by Brown (2003), "Moreover, absent face-to-face contact and ordinary non-verbal clues, even very mature students on the Internet demand more frequent interaction and reassurance in dialogue with their professors, an observation confirmed in student course evaluations" (p. 58). Brown (2003) went on to say that "teaching online is far more intense and absorbing from the instructor's point of view and, therefore, simultaneously more exhilarating and more exhausting mentally" (p. 59). In other words, constant feedback and interaction are musts for digital classrooms, which actually increases time spent working for the instructor.

Another myth about online classes is that impersonal relations are created among the teacher and students. The reality is that posting discussions online calls for students to think more carefully about the wording of their thoughts, and, in turn, the actual thoughts themselves. Brown (2003) claimed, "I believe this intimacy results from a sense of shared control and responsibility, commitment to collaboration and dialogue, and increased willingness to take risks in communications with others online" (p. 59). The dedication to effective communication by the instructor and participants diminishes the existence of impersonality.

The Business World: Myths about Telecommuting

With all of the technological innovations being made in secondary schools and higher education, the consideration of conducting business from home was inevitable. Telecommuting is on the rise and is creating a stir from those used to the typical nine o'clock to five o'clock workday. According to Dziak (2001), "If you regularly read or write reports at home, check email from a hotel room or the airport, or fax documents from somewhere other than your work office, you are telecommuting" (p. 143). Telecommuting is a task all too familiar to highly productive professionals but is accompanied by myths about its actual productivity.

One myth is that telecommuting is an easy way out of work and dealing with the work schedule. Several professionals believe that telecommuting allows for sleeping in, saving money only on childcare, and lessening the control of the employers over employees. Dzaik (2001) asserted, "The truth about telecommuting is that it can actually save employers thousands of dollars a year, due to space savings, productivity improvements and employee retention" (p. 143). Telecommuting builds trust and flexibility in the workplace, granting a proper balance between work and home life. As a result, telecommuting can increase employee output due to more comfortable work conditions. This modernization is not an everyday strategy; it lets an employee work from home about five times a month to complete projects. Dzaik (2001) supported telecommuting by saying the following:

> Most office-based telecommuters can expect to accomplish 30-50% more than they would at the office. For these individuals, working at home has the same effect on productivity as coming in to the office on a Saturday, when no one else is there; that is, it allows tasks to be accomplished with total concentration, and with greater speed. (p. 144)

It is important to note that the previously-mentioned benefits of telecommuting mainly apply to trusted employees, as self-discipline, adequate time management, and technology proficiency are all factors (Dziak, 2001).

Conclusion

A number of myths exist regarding technology and its effect on productivity in education and business. As the general world continues to advance in the technological aspect, so do the education and business worlds; this is to be expected. The true task comes in enlightening students and professionals about the possibilities of technology and both its positive and negative consequences. Awareness and proper implementation of programs are fundamental as digital environments are employed to make better gains.

References

Brown, B. (2003). Digital classrooms: Some myths about developing new educational programs using the Internet. *THE Journal (Technological Horizons in Education), 26*(5), 56-61.

Dziak, M. (2001). Debunking telecommuting myths. *Chemical Engineering, 108*(11), 143-149.

ETHICS

6

The Role of Ethics in Technological Innovation

JJ Asongu[6]

Corporate scandals that rocked the U.S. economy in the first half of this decade have pushed the issue of business ethics passed the confines of academia into the board rooms of most companies. This is especially true for companies at the leading edge of innovations as they must make serious ethical decisions about the development of their new technologies. However, before delving into the detailed role that ethics plays or should play in leading innovation, it would be fitting to explore the meaning of ethics.

---

[6] Januarius Jingwa (JJ) Asongu, Ph.D., is an adjunct professor of Business Communication at the University of Phoenix Online and also an adjunct professor of Web Design at Chattahoochee Technical College, Marietta, GA. In addition, he is the editor of the *Journal of Business and Public Policy* (JBPP), as well as a business consultant specialized in strategic management, marketing, and international business. He has published numerous books and articles on a variety of subjects and is actively involved in various social causes.

Defining Applied Ethics

Ethics can be a very interesting subject in that it touches almost every aspect of human endeavor and it is actually a very complex subject. This paper will attempt to elaborate on "applied ethics," one of the three primary areas of the broader field of ethics (which includes meta-ethics and normative ethics) because it has an implication on the "ethics of technology," and in today's wired economy, most innovations are technology driven.

Applied ethics is a subdiscipline of philosophy that attempts to apply ethical theory to real-life situations. It makes use of theories developed by meta-ethics and normative ethics, related and sometimes overlapping arenas in the field. A simple, but illustrative example is the issue of cloning. It can be treated as an applied ethical topic since it involves a specific type of controversial behavior. However, it is most often also treated under normative principles - such as the right to life - principles which are often litmus tests for determining the morality of that procedure. Even more interestingly, cloning can be discussed under meta-ethics because it deals with issues such as "where do rights come from?" and "what kind of beings have rights?"

The ethics of technology belongs to applied ethics since it is that subfield of ethics that addresses the ethical questions specific to the technology age. One of the pioneers in this field is the German philosopher, Hans Jonas (1985), who has addressed the social and ethical problems created by technology. The ethics of technology is rightly considered applied ethics because technology itself is incapable of possessing moral or ethical qualities. Technology is merely a tool for getting things done, so the ethics of technology refers instead to two basic subdivisions: the ethics involved in the development of new technology, and the ethical questions that are exacerbated by the ways in which technology extends or curtails the power of individuals.

Ethics in Innovative Processes

Dooley and O'Sullivan (2001) argue that one of the core competencies of a successful organization is its ability to successfully manage its adaptation, change, and innovation. One cannot successfully manage these elements without thinking about their ethical implications. Given that change is the essence of business growth (Clarke, 1994), it goes without saying that growth organizations must be very ethically conscious.

An excellent example of where innovation and ethics intersect is with the cloning debate. Although cloning and creating embryonic stem (ES) cells as a source of tissue for transplantation has been banned in many countries on ethical grounds, the jury is not yet out on that. Savulescu (2000), for instance, has called for the Australian government to remove the restrictions. He argues that every day, people die because there are insufficient tissues available for transplantation. For him, the development of cloning and ES cell line technologies offers real hope for developing better sources of tissues for transplantation. He states that these new technologies may mean that damaged tissue (e.g., after a stroke or heart attack) can be replaced with normal functioning tissue rather than scar tissue. Savulescu regrets the fact that research into "therapeutic cloning" and the development of ES cell lines is illegal in several States in Australia, and therefore calls for the review of legislation in order to allow destructive embryo research.

Savulescu (2000) insists that at least research should be allowed on spare embryos from assisted reproduction. He thinks there is a strong moral case for this position; arguing that the above reason is just one moral view (of several plausible ones) of the status of the embryo that precludes producing embryos for research. He says that those who argue against his position are mistaken, and so it is morally permissible to produce embryos for research into therapeutic cloning.

In this case, ethical considerations are playing a negative role in that they are limiting the innovative process. This is not bad in itself because not all innovations are good for the well being of society. This case also highlights the fact that ethics is not as simple as it might appear. Deciding between right and wrong, or good and bad, can actually be a very difficult and controversial thing with well intentioned moral philosophers taking opposite sides of the issue.

Conclusion

It would be interesting for researchers to continue the debate on the role of ethics in technological innovations. This paper made it clear that there is still a lot of controversy within the ethics of technology. However, it made it clear that ethical practices should form the basis of any organization's innovative process.

References

Clarke, L. (1994). *The essence of change.* London: Prentice Hall.

Dooley, L., & O'Sullivan, D. (2001). Structuring innovation: A conceptual model and implementation methodology. *Enterprise and Innovation Management Studies,2*(3), 177-194.

Jonas, H. (1985). *The imperative of responsibility: In search of ethics for the technological age.* Chicago, IL: University of Chicago Press.

Savulescu, J. (2000, August). The ethics of cloning and creating embryonic stem cells as a source of tissue for transplantation: Time to change the law in Australia. *Australian & New Zealand Journal of Medicine, 30*(4), 498-502.

7

The Integration of Technology in Education

Precious Prothro[7]

The world is changing constantly. Given this fact, what we know today will be outdated knowledge tomorrow. The one thing that is certain is that our children need to learn and grow in knowledge; for knowledge is power. The traditional way of learning is no longer the only way to learn because technologically enhanced ways of learning are readily available and accepted. There are times when new technology and innovative issues bring debate and controversy among traditionalists. There are ethical concerns that must be addressed when extending

---

[7] Precious Prothro has been teaching for three years and loves her profession. She received her BA in Middle Level Education with concentrations in Language Arts and Social Studies. She has a Masters of Education in Curriculum, and is currently undergoing studies at Argosy University in Atlanta towards a Doctor of Education degree. Ms. Prothro is also certified in Science, an area in which she teaches at Dutchtown Middle School in Henry County, GA. Her goal is to design curriculum and also to become a school administrator.

new technologies to our children. Technology and education can have a lasting relationship as long as ethical boundaries are not crossed.

What is Technology and Ethics?

Terri Kingsley (2007) quotes Janet Brown as saying that technology is a process that "usually begins with a design brief based on a problem we have that needs solving" (¶6). At Ohio State University (2007) there is a focused class that teaches students about technology education. Technology Education is the study of "how people alter their environment by making changes in the forms of materials and the societal problems which result from human use of technology" (¶1). Together technology and education can bring students to a higher order of thinking. Technology, as it relates to the mind, depends upon the thoughts and processes of an individual's outlook. Technology, as it relates to computers, deals with ethics related to the manner in which one chooses to filter through, or expose oneself through search engines, networks, and so on. Sociologist Raymond Baumhart of the Markkula Center for Applied Ethics (2007) says that ethics refer to "well based standards of right and wrong that prescribe what humans ought to do, usually in terms of rights, obligations, benefits to society, fairness, or specific virtues" (¶8).

Integrated Technology and Education

There are several models used for achieving higher level thinking. These models can be advanced for some students; however, they can be modified to fit all learning styles and levels. It is not always necessary to use models of that caliber because there are modern means by which critical thinking can be merged into education, and that is through technology. There are schools that start students, in the ninth and tenth grades, working on technical skills that may not land them in the traditional colleges. However, these students are using their minds to nurture or cultivate a skill that has developed from a technical thought, which can produce an ethical outcome. There are schools that focus on the direct development of the critical and technical development of the student. For example, Bergen County, in New Jersey, (2007) has a program that teaches in this vein. In particular, Bergen County Technical School Teterboro's campus offer classes in computer science, commercial art and graphic design, law and order, cosmetology and skin care, and so on. In short it is about reaching "excellence in education for all students," which is "going above and beyond to challenge and support each student to reach his or her full potential" (¶1).

Technology reaches heights far beyond our limited knowledge. In order to increase the knowledge our students obtain we must find a way to guide the students toward learning. Learning should not be forced, and there is a freedom of learning that must be explored. In the Code of Ethics of the Education Profession for the National Education Association (2006), goals of the profession are stated. "Essential to these goals is the protection of freedom to learn and to teach and the guarantee of equal educational opportunity for all" (¶1). The integration of technology, in all its vast forms, may be the path to help a student find his or her niche. We will not know if we do not allow them to explore.

Outside Involvement: Parents, Teachers, Community/Government, and the Student

Parents play a very vital role in a child's learning, and the involvement they share can make the difference in the child's success. Wikelund and Cotton (2001) conducted a study about the effects of parental involvement on student acievement. In this study it is stated that "educators frequently point out the critical role of the home and family environment in determining children's school success, and it appears that the earlier this influence is "harnessed," the greater the likelihood of higher student achievement" (¶12). The teacher's role can be challenging and can force the student to make a difference. The educator must request parental support, but before this can happen this need must be understood. An educator and a parent are a part of a four member team and each entity of that team has to support each other, and most importantly, the child. The method of teamwork may not always work, but roads must be traveled and explored to find the best paths for each student. The students will not always understand, but they will learn to appreciate it.

It is the job of the parents and the educators to "make" the government/ community and the community understand the importance of learning for the students—one child at a time. It is important that the community/government is involved in every decision that is made for our children, but the problem is that they are often far removed from the classroom. The educational, ethical, and technological decisions the team makes are in the team's hands; they must make the right choice. The future of our society lies in the education of our students, and each entity of this four member team has a very vital role to play. An incomplete job on the part of one team member does not mean failure for our students; however, completion and total involvement from all does lend a strong vibe of success. Technology and ethics will assist educators, parents, the community/ government, and our children and our students to grow by leaps and bounds. If everyone in this team remembers the goal, our children, then success will be the reward.

References

Bergen County Technical Schools. (2007). BCTS faculty. Retrieved September 30, 2007, from http://bcts.bergen.org/index.php?option=com_content&task=view&id=226&Itemid=693

Kingsley, T.A. (September 5, 2007). *Technology expos: What is technology?* Retrieved September 22, 2007, from http://www.abc.net.au/centralvic/stories/s2024605.htm?backyard

Markkula Center for Applied Ethics. (2007). *What is ethics?* Retrieved October 15, 2007, from Santa Clara University Web site: http://www.scu.edu/ethics/practicing/decision/whatisethics.html

National Education Association. (NEA). (2006). *Code of ethics of the education Profession.* Retrieved September 22, 2007, from http://www.nea.org/aboutnea/code.html

The Ohio State University. (May 2007). *What is Technology Education?* Retrieved September 22, 2007, from http://ed1.eng.ohio-state.edu/WhatisTE.html

Wikelund, K. R. & Cotton, K. (2001). Parent involvement in education. Retrieved September 30, 2007, from http://www.nwrel.org/scpd/sirs/3/cu6.html

8

Technology, Ethics, and Counseling

Kimberly Seheult[8]

As society develops and becomes faster paced, so does the need to get things accomplished quickly and proficiently. Time is of the essence and with the evolution of technological advances there seems to be more time to do other important things which allow for more growth and development. With email, text messaging, instant messaging, video conferencing and chat room cyber communication, we can send and receive messages instantly and globally. These advances in communication

---

[8] Kimberly Seheult is a counseling psychologist who is currently pursing a Doctor of Education degree at Argosy University, Atlanta. Originally from Trinidad in the Caribbean, she has lived in Georgia for 10 years. Although her background is in Psychology and Community Counseling, she has decided to branch into Education because she wants to be a part of change at the early stages of an individual's development. Her vision is to open a Holistic center for adolescents that will focus on individualism, self esteem, self actualization, behavior modification, and socialization.

have bridged the gap, connecting the world as we know it; geographical distance is no longer an unconstructive or disadvantaged factor, and things can happen instantly. These developments in communication are now being used as a medium between counseling and counseling education.

Cyber Communication and Counseling

Although cyber communication has not reinvented counseling, it can establish a more useful means of communication across therapeutic boundaries, such as distance. Some individuals may not have the necessary means of transportation or their area may not offer a range of therapeutic approaches to meet their individual needs (Lundberg, 2000). This would allow a more multicultural therapeutic choice which can reduce resistance in the therapeutic relationship (Elleven & Allen, 2004). One may argue that this profession is based primarily on face-to-face empathetic interpersonal skills. To channel a technological world and personal contact is of much concern to those in the counseling field.

There are various means of communicating other than the traditional face to face methods that have been warmly received including the use of the telephone in hotline suicide prevention, 24-hour mental health services, and mental health consultation (Reimer-Reiss, 2000). Now with the use of email, chat room (e.g., group therapy), and video conferencing, people less likely to visit the traditional clinics are able to utilize a professional without leaving the comfort of their homes. There is concern regarding whether or not this medium of communication would in fact be reliable. This medium that transfers the information can distort or change it in some way (Wilczenski, 2006). A message when sent via email or text cannot indicate the feelings that are being felt by the sender; many may not have the language skills to describe where they are emotionally at that particular moment. Cultural and language barriers have to be taken into consideration for an effective client-therapist relationship. Video conferences are a useful way of getting a more packaged deal. One is able to communicate over vast distances, and the visual aid is there to facilitate complete communication. A counselor is better able to use appropriate techniques when nonverbal communication is also present in the session. This is the one technological medium which most resembles counseling in its true form (Reimer-Reiss, 2000).

Ethical Considerations

The greatest challenge to date is cyber counseling compromising the therapeutic relationship. Cyber counseling is a great concern to the ethical premise to which psychotherapy and philosophy holds itself. To ensure this type of therapeutic intervention does not test the bounds of sensitive issues, confidentiality and informed consent, the use of this medium must be taught by professionals who understand the intrinsic value of what it means to execute cyber counseling. These technological services would need to be constantly monitored and scrutinized by the organizations that implement them and by the professional body which is psychotherapy (Wilcenzenski, 2006).

Unconventional Alternatives

Technology will not wait on any person or profession. The inevitable is to find a way as a professional to bridge your own cyber therapy gap. Those counselors slow to this new info structure maybe left behind (Elleven & Allen, 2004). Technology does not have to take away the humanistic side of this profession because the alternatives are possible. A blended style, such as online and personal contact, can be an effective choice without losing the core of psychotherapy. If a client is able to communicate effectively and there has been an initial stage of rapport, the therapist can consider incorporating a virtual feature. This can include online journaling, specific chat time, video conferencing, group work, and other therapeutic assignments online. For this to be practical and remain confidential the therapist must have a specific site used only for therapeutic intervention where all conversations and assignments are completed discreetly. To guarantee confidentiality, the used site must have password access and should be monitored by the state licensing board. This would also indicate client development and monitor appropriate procedure. This can aid in incidents of malpractice. It is imperative to note, the importance of personal communication that makes for effective therapy. Virtual contact would not diminish this process but may help develop a stronger commitment and openness to the therapeutic process.

Conclusion

It is important to understand the positive and negative property to such a powerful and useful medium. The ability to reach the needs of people in all areas of the world, to give clients another form of communicating, and to reduce the socioeconomic barriers should be embraced. Therapy is a means to encourage individuals to take control of their lives. Cyber counseling is an instrument that can be used to promote a client's internal want and drive. This is not the only recommended means of counseling but should be considered when face-to-face contact is impractical or if the needs of the therapeutic relationship could be met using this medium. This should be a choice of both the therapist and the client, with the therapist using his or her professional discretion.

References

Elleven, R. K., & Allen, J. (2004). Applying technology to online counseling: Suggestions for the beginning e-therapist. *Journal of Instructional Psychology, 31,* 223-226.

Lundberg, D. J. (2000). Integrating on-line technology into counseling curricula: Emerging humanistic factors. *Journal of Humanistic Counseling, Education & Development, (38),* 142-151.

Riemer-Reiss, M.L. (2000). Utilizing distance technology for mental health counseling. *Mental Health Counseling. (22),* 189-204.

Wilczenski, F. L. (2006). Cyber-communication: Finding its place in school counseling practice, education, and professional development. *Professional School Counseling, 9,* 327-331.

MONEY

9

Funding of Technology in Schools

Mary Frances Howard[9]

There are a myriad of mixed emotions about educational funding. Schools districts across the country are having a dilemma about how to disperse funds and where they are needed the most. Weeks (2004) contends, "The expanding capabilities of telecommunication technologies have combined with strong movements favoring public school choice, charter schools, and home schooling, along with the stimulus of technology-funding initiatives and the No Child Left Behind Act" (U.S. Department of Education, 2001, p. 2). According to the

---

[9] Mary Frances Howard is a Language Arts teacher in Savannah, GA. She is an experienced teacher who was honored with the County Teacher of the Year Award for 2006-2007. Among her academic credentials are a Specialist degree in Educational Leadership, and in addition she is Certified in Administration. Ms. Howard is currently pursuing a doctorate in Instructional Leadership at Argosy University, Atlanta.

article, "An Educators Guide to Funding Issues" (Byrom, 1998),

> Meeting the needs of schools, teachers, and students in this technological age does not come without a price tag. [Therefore], schools and communities must make choices about funds to provide the hardware, software, infrastructure and support to be able to use and maintain technology. (p. 1)

Funding Strategies

There are at least three challenges that have to be addressed to satisfy the funding issues of technology in schools. Sun (1996) explains,

> First ways must be found to fund relatively high level of capital expenditures needed to install school technology systems. [Following this], funds to sustain school's annual operating costs must be identified. Finally, funds must be secured to regularly retire and replenish portions of the system to keep it modern. (p. 1)

In an effort to keep up with technology's ever-changing facets, there has to be stringent and effective methods to obtain the much needed revenue to secure the sustainability of technology.

Affordable Technology for All School Districts

In the past, many school districts have approached instructional technology as somewhat of an extracurricular activity. In preparing our students for the 21st century, that idea is null and void. Instructional technology is a vital part of the curricula globally. We are in a technical age where school districts would be remiss if they were not utilizing technology to its fullest capacity. It is imperative that all schools have the opportunity to have affordable and effective technology by which to teach our students. Sun (1996) states, "Although schools have budget lines for staff salaries, transportation, building and facilities maintenance, textbooks, and other essentials, few schools allocate regular funding for the purchase, maintenance, upgrading, support, and professional development to new technologies" (p. 1). Many districts rely on grants, special bond issues, and charitable donations to help with affordable technology. Nevertheless, there seems to be more avenues districts can rely on to ensure appropriate funding for technology.

Federal Contribution

Although the federal government contends it is for assisting our students in becoming academically proficient, it is not as benevolent when it comes to providing revenue for educational technology. According to the article, "Funding Strategies for the Technology Plan" (Vojtek, 1997), "One of the most productive uses for federal funds, particularly at the state level, is in planning for the development of technology in schools and the schools' connectivity to the information superhighway" (p. 3).

> Moreover, the federal government could provide tax incentives to help fund school technology. [On the contrary], tax incentives given directly to businesses for donations of used computers as provided for in the Tax Relief Act of 1997 (H.R.2014), may prove costly to the government and could result in donations of equipment that is not needed. To avoid these problems, a voucher tax credit scheme could be used in which the value of the income tax credit

> is given to the states in the form of vouchers. These vouchers could then be used by the states to purchase technology. (Sun, 1996, p. 2)

State Contribution

"Although the federal government will have important roles in implementing the national educational technology plan, the responsibility for adopting coherent and coordinated funding strategies lies with state governments" (Weeks, 2004, p. 3). It is much easier for the state to allocate funds directly to the districts to ensure high-quality educational technology for all students. Unfortunately, the latter is the exception instead of the rule because of how many states are forced to allocate money into more needed areas of education. As a result,

> School districts that do not use the [allocated] money immediately for technology could be given one of two options. Either they could put the funds into an interest-bearing account to be spent at some future time on technology, or [as mentioned before], they could spend the money on other things. (U.S. Department of Education, 1997, p. 3)

Conclusion

In brief, the issue of funding technology for education will continue to be a hot topic on the minds of teachers, administrators, parents, and stakeholders. We have to contend that we are in a globally technical society, and if we are to compete in this cyberspace world of the Internet, Blogs, and MySpace, adequate revenue has to be provided.

References

Byrom, E. (2007). *An educator's guide to funding issues.* Retrieved September 23, 2007, from http://www.ncrel.org/sdrs/areas/issue/methods/technlgy/te41k22/htm

Sun, L. (1996). *Funding strategies for school districts to promote engaged learning through technology.* Unpublished manuscript.

U.S. Department of Education. (1997). *Technology innovation challenge grants.* Retrieved September 23, 2007, from http://www.ed.gov/Technology/challenge

U.S. Department of Education. (2001). *No child left behind.* Retrieved September 23, 2007, from http://www.ed.gov/print/nclb/index.html

U. S. Department of Education. (2007). *Investing in school technology: Strategies to meet the funding challenge.* Retrieved September 15, 2007, from http://www. ed.gov/pubs/techinvest/chap3.html

Vojtek, R. (1997). Using technology to emancipate time. *Journal of Staff Development*, 6, (97).

Weeks, D. (2007). The search for funding. *NWREL, 20* (3).

Made in the USA
Monee, IL
08 November 2026

67377454R00044